PLAYGROUND EDUCATION

CREATED and WRITTEN by

R. TOBIAS PITTMAN

THE EDUCATIONAL PANDA

PUGSLEY PANDA

ILLUSTRATIONS by

William Chislum

EDITED by

Gloria Miles

Playground Education and Entertainment
www.pugsleypanda.com | email: pugsleypanda@gmail.com

Hello Pugsley,
do you have an umbrella?
I think it is going to rain.
Hello Renee,
I do not have an umbrella.
May I share your
umbrella with you?

Hello everyone, did someone call my name?
Hello Science Scott, we were talking about the rain. You know all about science, maybe you can explain to us where rain comes from.

Listen carefully,
I will explain where rain
comes from.
Great,
I am excited!

Actually all the water we see and use including rain, has been around for millions of years. The water keeps going up to the clouds and back to the ground, up to the clouds and back to the ground. This is called "The Water Cycle".
Wow, that sounds interesting. How does that work?

The same thing happens at home, when you pour a glass of cold water on a hot day. Water forms on the outside of the glass.
Some people think the water somehow magically leaked through the glass, but it actually came from the air.
Water vapor in the warm air, turns back into liquid when it touches the cold glass.
This is terrific!

Precipitation is the next step in the water cycle. Precipitation happens when so much water has condensed, that the air cannot hold it anymore. The clouds get heavy and water falls back to the earth in the form of rain, hail, sleet or snow.
Wow, that's incredible! I need my umbrella again.

Collection is the next step. When water falls back to earth as precipitation it may fall back in the oceans, lakes or rivers where the water is collected.
This is awesome, Science Scott!
If the water ends up on land, it can soak into the earth and become part of the ground water that plants and animals use to drink. This is where the cycle starts all over again.

Do you guys understand the water cycle now?
The explanation of the water cycle was very interesting and educational, Science Scott.
How can I remember the steps?

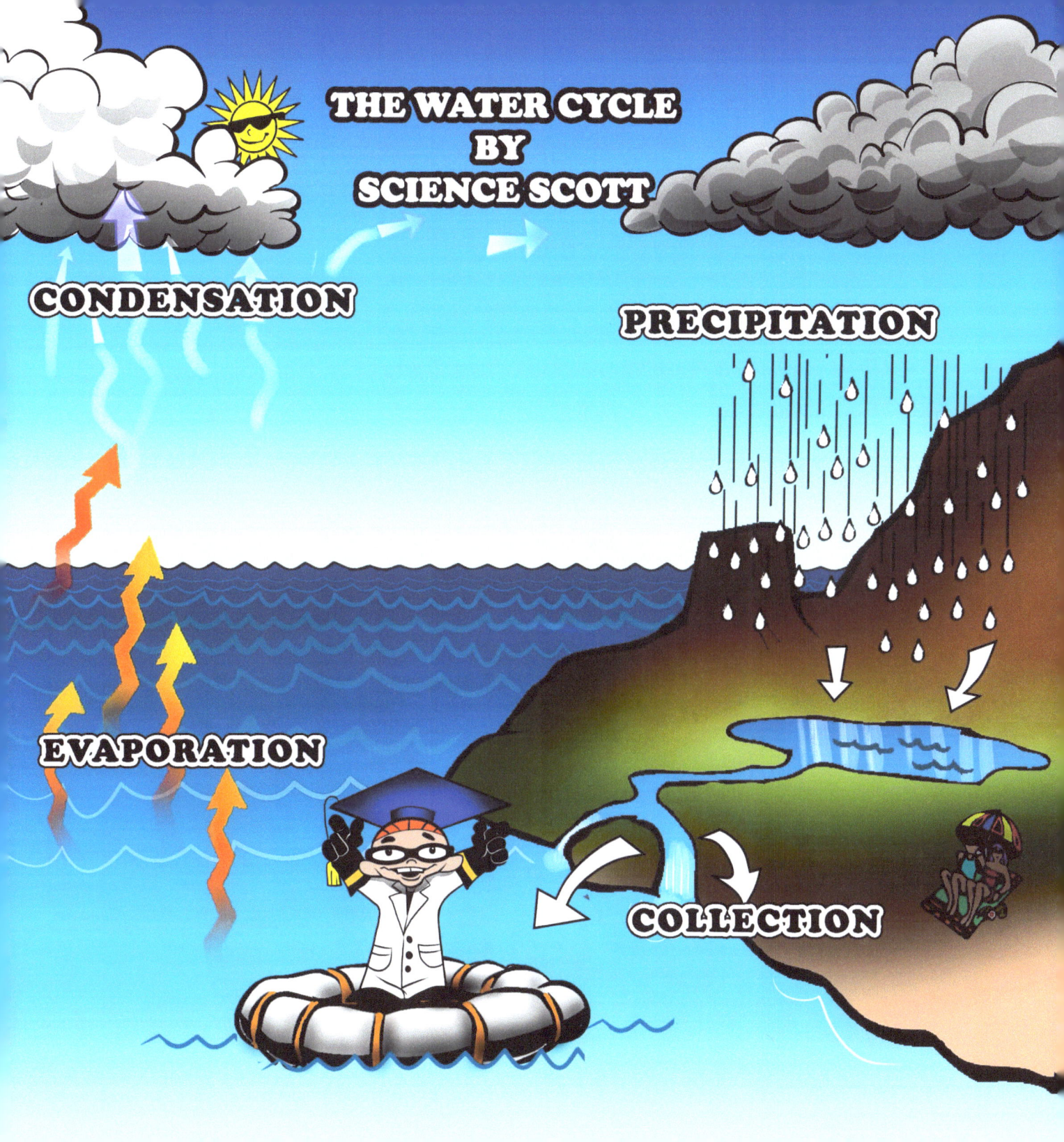

ENJOY THESE OTHER CHILDREN'S BOOKS FROM

PLAYGROUND EDUCATION

"No Bullying"
By

Safety Sam

"Wellington Hill Playground"
By
The Educational Kidz

"Famous Landmarks in The United States"
By

Geography Gerald

"Seasonal Safety Tips for Children"
By

Safety Sam

PUGSLEY PANDA